I0494356

ROBOTICS

by Oksen Babakhanian

RoseDog Books
PITTSBURGH, PENNSYLVANIA 15238

The contents of this work including, but not limited to, the accuracy of events, people, and places depicted; opinions expressed; permission to use previously published materials included; and any advice given or actions advocated are solely the responsibility of the author, who assumes all liability for said work and indemnifies the publisher against any claims stemming from publication of the work.

All Rights Reserved

Copyright © 2021 by Oksen Babakhanian

No part of this book may be reproduced or transmitted, downloaded, distributed, reverse engineered, or stored in or introduced into any information storage and retrieval system, in any form or by any means, including photocopying and recording, whether electronic or mechanical, now known or hereinafter invented without permission in writing from the publisher.

RoseDog Books
585 Alpha Drive, Suite 103
Pittsburgh, PA 15238
Visit our website at *www.rosedogbookstore.com*

ISBN: 978-1-63764-683-0
eISBN: 978-1-63764-723-3

Oksen Babkhanian
In collaboration with Ailin Babakhanian

ROBOTICS

This book is about the new age that we are on the verge of it, and soon it would be an age that is totally different than human beings have experienced before. The COVID-19 pandemic shows a glimpse of what the future is going to look like and how people are going to stay and conduct their life and business affairs from inside their homes. The advent of new and highly technical and progressed robots that do most of the jobs for human beings, creates a new world where people do not need to work and will be transferred to different locations with automated robot-driven vehicles. In such a situation, we are here to examine how people will be surviving and what the social agenda would likely be. We think in the new age the social life will be of a life in an educated socialism society.

CONTENTS

PREFACE

In our opinion, it is the time to thoroughly examine the social life in the near future when the robots will take over most of the jobs that humans are doing, including: creation of goods, warehousing, transportation, driverless cars and airplanes and boats, automated kitchens and restaurants without chefs or waiters, Army with robot soldiers, hospitals with robot doctors, surgeons and staff. In such a society, how people and government would interact and how the life will be for the citizens. The best scenario will be to have a new educated socialistic life—meaning that, since people are staying home most of the time and jobs are done by robots, the government has to provide a universal standard salary for people to live in comfort with free basic housing and healthcare.

In this type of society, the emphasis will be the citizen himself to have the best education and emphases on humanity, arts, sports, and music rather than prepare students for mundane works. Those who wish, can get a highly technical and sophisticated education in robotics and other fields to be part of those new age scientists and business innovators. The new businesses will have exceedingly small staffing to manage the projects because everything else will be done by robots.

EDUCATED SOCIALISM

When Bernie Sanders talked about some socialistic agenda, lots of people called him, socialist because they did not understand his overall message. They heard the word socialist and being die-hard capitalists, they all protested and did not vote for him on his election bid. What he was talking about was only a small portion of socialistic agenda. In this age of robotics, we are going to listen much more about socialistic life because due to an increase in capacity and capabilities of robots, we are going to have a more of socialistic life in the US and most other developed countries.

Now, the educated socialism is somehow different from the original socialism that many countries in Europe practice. Most of all, the educated socialism means that we are going to have the basic promise of socialism with emphases, being on having a "Universal Basic Income" so that every family will live in a middle-class lifestyle with enough income to cover habilitation, education, and healthcare. After which, people who want to excel and people who have higher education will tend to get higher-earning jobs and run bigger businesses such as Amazon, eBay, Google, etc. Educated world in educated socialism is for people to have enough resources to live in basic comforts so that they are free from mundane bills and expenses; then they can concentrate on getting higher education and higher degrees in businesses, engineering, computer science, chemistry, physics, biotics, etc.

By doing this, those highly educated businessmen and entrepreneurs, doctors and such, will be able to create huge money out of their businesses, or to be much more productive in fields of medicine and treatment of all kinds of pandemics or other virus- or bacteria-originated sicknesses. With those two concepts hand in hand, human beings will have a much more comfortable life with fewer daily headaches and sicknesses so they can concentrate on higher education, very progressive and technical subjects and work on mega businesses to make millions in revenue—a good chunk of which will be taxed. With those taxes, the government will pay the "Universal Basic Income" and provide highly sophisticated and superior roads, buildings, schools, hospitals, army, and so on.

COVID-19 PANDEMIC

Because of the COVID-19 pandemic, the lifestyle in most of the world has changed dramatically and started the new robotic age concept of living. With the COVID-19 pandemic, people had to stay home and work from home by use of new technologies such as Zoom conferencing and use of computers connected to their offices, businesses, and classrooms. By doing this, people stayed home instead of driving their cars to get to the office or schools and run their businesses remotely by their computers, cell phones, FaceTime, and other means of communications. Due to this phenomenon alone, the freeway and street traffic jams are gone, gas stations are idol, lots of businesses are out of business and lots of workers are jobless.

People have got used to ordering their food, their clothing, their furniture, and appliances all online. The online orders are so many that because of it, new companies such as Amazon were expanded to become the highest grossing company in the world and create huge warehouses which are run by few employees and millions of robots. Therefore, with this pandemic lifestyle, people tasted the coming lifestyle in the age of robotics. The lifestyle is not normal anymore and it is a "New Normal" with its new requirement and style of living which is much different and unique than the lifestyles us humans have had previously for centuries.

The pandemic teaches us to be more productive at home with use of existing or new technologies without wasting lots of time and money on transportation and transfer of goods. we are learning to stay home, to exercise at home without going to gyms, cook or order foods rather than going to restaurants, and most of all, learning to deal with family members, all in the house with their unique needs and feelings. We start to learn how to create our own space in a small home when there is no space. That is why all kids and adults, Mom and Dad side by side, have their cell phones with social media, news, emails at the same time without recognizing the person near them. We all have become zombies in our own home with cell phones in our hands, using Zoom and FaceTime to do our business and talk to friends and waste our time on social media, with recent video clips of all kinds, new photos and stories, and then we lose the track of time, meanwhile we did not notice the person sitting beside or across from us.

WAR IN NAGORNO KARABAKH (ARTSAKH)

Nagorno Karabakh is a small (4,400 sq kilometers), enclave situated in Azerbaijan, on eastern boarder of Armenia and populated by Armenians. This enclave was part of Armenia for centuries but unfortunately in the time of Soviet Union, Josef Stalin took that enclave from Armenia and handed it over to Azerbaijan in 1921. Then, after the collapse of the Soviet Union in 1991, Nagorno Karabakh autonomous region sought independence from Azerbaijan to join her motherland of Armenia. There was a war, and finally there was a truce in 1994 that lasted 'til September 27, 2020. Then, Azerbaijan, with Turkey's help and use of Arab jihadists, attacked Karabakh enclave for recapturing and doing ethnic cleansing from Armenians by bombing cities including churches, hospitals, schools, and civilian residences.

This war was one of the new types of war where new technology of drones was used. Azerbaijan used their unmind drones of highly technical sort from Israel and Turkey. With the help of radar system from F16 war planes, they could pinpoint exact location to bombard targets below. The Armenian defense army with highly trained and superior technological advance education and most of all their dedication on sa-ving the homeland, they did their best to do a conventional war but at the end they were the losers because they did not foresee the coming and use of new drone technology. Those drones in the sky were target-ing Armenian soldiers, tanks, and other weapons and essential buildings or other targets on civilians.

The Armenian army not having the proper technology to download those drones finally had to end the war with big losses. The drones are unmanned new technology that are very sophisticated and run on ra-dars. They can do the job of heavy army airplanes, and they are versatile and cheap to make and are used on wars without using soldiers on the ground. A country like Turkey, who borrows technology and parts from other countries, makes thousands of those small drones, which it used in both Syria and Karabakh wars and were victorious in both. As a matter of fact, drones are robot-driven new technology used in new wars; it shows that a lack of having this new robotic technology, coun-tries will most probably lose the conventional wars. In the future, most of the wars will be conducted with robot soldiers, drones, and other new technological advanced weapons.

DRIVERLESS CARS

One of the main aspects of our future robotics life will be having driverless cars. Those cars will be driven by robotics/computers without help of any human being. Our taxis and our own cars will be of these sorts, and there is no need to hire drivers, drive yourself, or get a taxi with a person to drive. This phenomenon will be a huge change in our society as far as transportation goes. In one hand, it will eliminate millions of jobs for taxi, truck, and all other drivers; in the other hand, families who have three or more cars in their driveway or garages would not need to keep them all because all they need to do is have one driverless car which will take everyone to the different locations, being jobs or school.

By having those driverless cars and the elimination of lots of family cars, the traffic will be minimum, the accidents will be reduced, and imagine all those idol gas stations, body shops, and mechanic shops. This aspect also will eliminate lots of jobs and make people jobless. In such a situation, lots of people will be jobless and stay home. This situation will cause governments to subsidize those families either creating more welfare recipients or to have "Universal Basic Income." To get this universal basic income, large companies need to pay much bigger taxes so that government will be able to pay such an income to families.

There is also another aspect of life associated with this stay-home condition that are boredom, depression, overuse of alcoholic and drugs, fights in the family, and break down of society, and work habits as we know for millions of people. Therefore, companies or governments need to think of other amusements, sporting, entertainments, and other activities to fulfill the time and get people to be engaged in an activity and not to be bored. Also, participating in Zoom class to get more education. Involvement will be an encouragement program for society.

Due to driverless cars, families do not need to have three or four cars with payments, insurances, gas money, and all the repair work involved. This will be a huge saving for family budgets. Regarding the economy of the country, this will hurt the economy because of the reduction of millions of cars in streets and all the associated economic benefits that were due because of having many cars driven by drivers of all ages. Roads will be less driven, so there would be no need to maintain them more often or create larger roads and bridges. Lots of companies, such as insurance companies, repair shops, body shops, driver's education, will be eliminated or reduced. Government will not get any more income from issuing traffic tickets. Police stations will be reducing staff because they do not need to check the streets for traffic violations.

EDUCATION SYSTEM

In the age of robotics, the whole system of education will be changed. As we saw in the pandemic, people stayed home and used Zoom classes. This aspect of life will continue in the future by having much better teachers who are robots and will teach us at our homes as private tutors or using Zoom or online classes with robot teachers. In normal school settings, instead of human teachers, there will be many more robot teachers who can do much more than the regular teachers such as they would teach, be the doctor at the same time, do the entertaining of students, or providing food for them.

Having robots as teachers will eliminates jobs for lots of teachers. Robots in classes will be much more effective and versatile and entertaining with much more knowledge than regular teachers would have. Also, future robot teachers will be able to analyze kids just by looking at their eyes and movements, take their temperature, or guess on their behavior as far as fears, anger, depression, or other behavioral issues. So many of kids nowadays are autistic, and they need special treatment and teachers. This will be ideal to have robot teachers who can help them fully, by having the proper knowledge and understanding their situation and behavior. In future society, the education system will change totally with having robots as teachers and scientists and doctors.

Lots of jobs will be eliminated, so there would not be any need to teach those subjects to students; also, lots of general knowledge classes in school will be eliminated due to in-house education by Google and all those video-lecture teaching systems. Even some highly sophisticated degrees for scientists and doctors will be eliminated because the robots will take over those projects and research. Robots will be working in labs, and they are the ones who would create new systems, medicines, vaccines, vitamins etc.

It is hard for us to imagine a future society where there are no human doctors, scientists, engineers, architects and even the lay people to do the mundane things because robots will do all of the above and even more. Therefore, our education system will be totally different or will be simplified for human use only and all the sophisticated education will be done by robots. One should wonder if we need any sophisticated education anymore for human beings because all those are done by robots and humans can have the lifestyle of centuries before with minimum education and lots of time to enjoy basic life.

HOSPITALS AND HEALTH SYSTEM

In the future, hospitals and health systems will totally be revolution-ized by robots being doctors and administrative staff and executive body for hospitals; everything will be different than we use to know today. When a patient goes to a hospital, even if so because most of the time, hospitals will come home in the form of a robot doing all the tests, checkups, MRIs and X-Rays and other procedures right on the spot. Minor surgeries will be done at home. Only very sophisti-cated surgeries and procedures will be performed in hospitals, there-fore, there would not be a need for many hospitals or hospital compounds with multitudes of rooms with thousands of patients. As a result, we will see an economic downturn and lots of layoffs for doc-tors, nurses, and hospital staff and all the maintenance workers. Also, due to this phenomenon the hospital built will be few and a small ver-sion of existing mega hospitals.

The health system, will also be changed by robots in such that most of the procedures and medicine will be changed and the drugs will have multiple effects so we do not need to take many medicines every day. They will be much more potent, and it will include also proper vacci-nation, enough enzymes to cure most of the nagging and long-term diseases such as diabetes, rheumatism, etc. People, in having this proper medication with the exact dosage and having the prevention from most of the diseases, will be much healthier and live a longer life.

And this is another issue because the minimum living age will change from eighties or nineties to maybe 180s or 190s. Then the whole lifestyle and families will be changing… there will be lots of older people around and not too many kids or youngsters. The fabric of society will change to a totally different one with new norms of living and new political systems to deal with phenomenon. With a change in the minimum living age, one would wonder if those over 100 years old would have any productive life or will be living pro bono.

Another wonderful aspect of having robots in the health system is that the world would be much better protected from all kinds of diseases because the same group of doctors—meaning—robots will be all over the world to treat all kinds of diseases such as malaria, chicken pox, and so on. There would not be much difference between rich and poor nations as far as their health system goes. There would not be so many deaths, especially in poor countries. This phenomenon will create another issue, and that is overpopulation with humans not dying that easily and living much longer. This overpopulation will create issues with providing enough food and other essentials for world population. To solve this overpopulation, there will be either more wars for limited resources or robots trying to conquer other planets for human living. No wonder, Mr. Elon Musk is not only dreaming about repopulating Mars but working on his rockets to be able to take human beings to Mars.

DOCTORS

Robot doctors with their super knowledge will do miracles for their patients in all situations and will cure more and more people. With their super knowledge, those doctors will eliminate most of the common sicknesses such as colds, flus, hepatitis, diabetes, and other sicknesses caused by different viruses, bacteria, etc. Worldwide, developed, or undeveloped, being poor or rich countries will have the same type of robot doctors, therefore, those doctors will eliminate most of the diseases which were killing millions of kids and grown-ups in especially poorer countries.

The robot doctors will not be as we imagine human-like doctor but could be computers who are inserted in bathrooms and kitchens, so whenever households use the bathrooms right then. their urine and fecal samples will be tested for any diseases and right then the proper medicine will be diagnosed on the spot. Also, while you are measuring your weight on the scale at the same time your blood and heart rate will be tested for other related diseases. When you are sleeping, the bed doctor will check all your breathing, mind, temperature, pressure points, and your sleeping habits and record your dreams to scan and evaluate your physical and mental health. In case you want this robot, the doctor will give you the interpretation and future of your dreams.

As we can see, your house will be your doctor too, so you do not have to make an appointment to drive in a traffic jam to see your doctor with your insurance card. In this situation, the people will get sick much less than before, and if they get sick, they will be immediately diagnosed and proper medications will be assigned and will be delivered by driverless cars to your home immediately without waiting for weeks to get all these done. In case of emergencies, or minor surgeries, robot doctors will fly to you with drones immediately and perform all the necessary rehabilitation needs and surgeries.

PHARMACY

In today's society, the pharmacy and pharmacist have a huge impact on the people and business. Millions of dollars are spent on different kinds of medicines being prescribed by doctors or over the counter medicines and vitamins. The use of vitamins has become so much more common that almost everyone uses many of those daily, which has a big price tag for family budget. Also, doctors prescribe drugs made by those pharmaceutical companies, which are useful, but the dosages are so small that people need to take those drugs daily to get required help and wellness.

It has become the norm that most people take five, six, or more pills daily for different health problems or just to enhance their health or to prevent coming sicknesses. In the robotic age, when robots daily will check the health condition of the person and would prescribe the exact pill, then there would not be this huge business of selling a multitude drugs with huge price tags. This in turn will eliminate millions of jobs, workers, pharmacists, delivery persons who deliver those drugs to elderly and others. As we can see, in this new age again, jobs will be disappearing, and workers are going to be jobless.

HEALTH INSURANCE

Today, health insurance is one of the costliest services for governments and society. People pay almost 10 to 15 percent of their income to health insurance companies to cover their general health issues; also, they are out-of-pocket fees on top of minimum health insurance payments. In the future, in the age of robotics, there would not be any needs for having health insurance because the houses already have most of the health-related technology and diagnoses built inside. There is no need to pay extra to insurance companies to provide you minimum health insurance. Also, since the need to use hospitals is so small, then the cost to use that service will be also at a minimum and people or the government will pay for that.

Although, this aspect of health-cost reduction is good for families, but it will also create unemployment for millions of health insurance companies and their employees. This is where, again, "Universal Basic Income" will be necessary to pay those unemployed workers. Due to the heavy cost of health insurance, millions of people did not have insurance, therefore, they got sick more often or they waited a long time to go to the doctor, thus, making their simple sickness become much more critical and costly and staying in the hospital much longer and much more expensive. By having the robotics take over the health system, we will also eliminate those costly situations of people not using proper medications and hospitalization.

DELIVERY (POSTAL, TRUCKING, AND SMALL ITEMS)

In age of robotics, all delivery systems such as, postal, trucking, and others will be carried out by robots. Robots in the shape of driverless cars and trucks or drones. Drones will deliver most of the small items door-to-door to the houses. Since they are light and agile, they can travel all around the town to different houses and deliver all kinds of food and mail as well. Driverless trucks will deliver the bigger items such as furniture, construction materials etc. in the process, lots of truck drivers, postal employees, and all others involved in delivery systems will be laid off because robots will replace them.

Also due to use of drones, there will be minimum traffic in streets, so it will eliminate traffic congestion, accidents, and most of all the city authorities or states do not need to spend millions of dollars to create new roads and infrastructures. The decrease in the cost and spending will be welcomed by governments so they can spend the additional money on needed projects. In the future, the drones will be so advanced that they can communicate with each other, make the delivery system very efficient and less costly.

This phenomenon alone will be a huge revolution in the delivery system for goods and other essential items. Items will be delivered not in days or weeks but in hours, especially in ordering food from restaurants even ordering from other cities for special dishes will be common. This will enhance the livelihood of people, but on the other hand will create millions of unemployment for workers.

RESTAURANTS

In the near future, when the robots will kill the chefs and waiters and waitresses, they will run the show by themselves. They will take the orders not only from inside customers but from all over town and even neighborhood towns. They will prepare the food per exact order received and delivered to tables to customers and deliver via drones all around the town in a few minutes or hours. The prepared food would include all the necessary information as far as the calories, vitamins, fat, sugar, and so on. Each food will be prepared individually for the customer's unique order at No Extra Charge! For big parties, such as birthdays, showers, engagements, weddings etc., the catering will be handled by robots, the valet parking will be handled by robots as well.

Although, there would not be too many cars because the guests will arrive in driverless cars. In such a scenario, one can imagine the job loss for millions of restaurant workers, chefs, and valet parking workers.

As we can see all that unemployment will make huge changes in the way governments are running, and those unemployed workers are getting paid by the government. Now how will those governments budget such a huge unemployment wage? That is when the suggested "Universal Income" comes into the picture that we will discuss in another topic. In future restaurants, robots will deliver food in many ways as in circus, with car, with horses, with hands coming from sky, with dancing robots, robots playing with foods, tossing up and down, making funny faces, and singing or telling you jokes while serving the food. It could be an ideal place to enjoy the food and get entertained as well.

AMAZON

Most people are familiar with Amazon corporation, especially because of the pandemic, everyone stays home and orders their needed items online from their favorite places by just a click on their websites. Amazon has huge warehouses that store millions of items and goods from different manufactures and sells them online. Those warehouses are mostly manned by robots. Those robotic machines sort the items, they put them on the shelves, and when the order is placed, they pick up the right item from the right shelf, and deliver it to the right truck so the truck driver can drive the item to the order destination. Amazon became an example of future sales and delivery of all kinds of goods. It became a millionaire overnight and although it has lots of employees, they eliminated thousands of warehouse employees by using robots instead.

Amazon soon will eliminate thousands more employees and use robots to replace them. The robots do not need sick leaves, maternity leaves, overtime payments, etc. Therefore, robots create lots of savings and reduce costs for the company, and they expedite the delivery of goods to customers. This phenomenon has made Amazon a billion-dollar company and its owner, Mr. Jeff Bezos the richest person in the world. What are the taxes Mr. Bezos or his company pays to the government?! In the future, companies such as Amazon, that create billions of dollars of income should pay for the "Universal Minimum Income," which is used to pay unemployment wages to those millions of workers laid off from their jobs.

FARMING

In the near future, farming will be done exclusively by robots. They will run the machinery, test the soil, add all the required pesticides and fertilizers to the soil, and the exact amount of required water for that certain crops. Therefore, the product will be exactly what companies and consumers are asking for. If a company asks for organic food, then the robot farmers will produce organic food as requested. Imagine, with the proper fertilizers and pesticides and watering system the efficiency will be multiplied, and the crops will be much better and healthier for consumption. Those robot farmers will determine the type of food needed to eliminate lots of sicknesses, or by some additions of certain chemicals they can enhance the quality of that product.

Also, robots as a researcher themselves, will test the final farm products for their quality, safety, disease prevention, and proper number of vitamins or other required enhancements. The final products will be measured and delivered to customers on time, per the specifications they receive from those customers. Since robots are in charge, they will check the crop quality day and night, test for water or other enhancement requirements on a daily basis by picking up soil samples and examining it right on the spot with their robotic knowledge about all kinds of disease and required fertilizer type and exact amount of water required for that certain type of crop.

This is like having your own doctor checking you twenty-four hours a day, seven days a week for quality and specification of the products. Those robots will prepare the farm product in specified packaging for delivery day or night. Therefore, the customer will get fresh products with extremely high quality and great taste and free from any poisons and harmful substances.

BACK TO BASICS

We used to listen to the stories of our grandparents when they used to live in small towns or in farms in villages, quiet days, having family dinners with all family members, sitting around fire at cold nights with shining stars above—the elders told stories and youngsters listened with wide eyes. In the near future when the robots are at work and do all the hard works preparing food, do the delivery and make the production of goods, working at warehouses, run the driverless cars, all of these and more will create millions of unemployment for common people. Those unemployed people need to survive by help of government or as we previously mentioned, by "Universal Minimum Income."

In one way, those people will have more time in their hands and possibly families could be together more and have those quiet dinners as they were told by their grandparents. However, in reality as we have seen recently in pandemic, families will stay at home but everyone, including children and elders, will be consumed by their cell phones and we do not see any sitting around together for having family dinner. As a result, individuals by use of those cell phones will be separated more not physically but emotionally from each other because everyone having their own taste on the social media and would tent to immerse themselves in to their hobbies and act more like zombies than those cheerful family members when gathered together.

In near future since people stay home and do not have any work to attend, if they are encouraged by what we called "Educated socialism," they would spend their time to get more knowledge, engage in physical activities, exercise, play many different games, such as basketball, football, tennis, etc. Also, they will participate in music, ballet, and arts. Due to being very educated and advance in human thinking, they will be more polite, caring, and loving members of their society. People will have more quality time to enhance themselves for the higher calling and be more religious attending churches. Also, the morality of the people will increase, and they will see all human types as one united nation not separating them by their colors, religion, or the language spoken. We imagine a world that is quieter with less stress, with people contained and cordial toward each other.

There will not be any wars any more, governments will not spend billions of dollars on armaments and preparation of wars because since robots are taking care of people and providing their food, clothing and shelters there would be no need to fight over those raw materials. The world will be quieter and will be a lovely place to live in. Finally, people will have more time to travel more not only within their cities and countries but also, other countries too. Being a world traveler, getting introduced to many cultures, people will tend to be more mature and understanding of each other knowing that human beings are all one with the same feelings and aspirations therefore, rather than trying to humiliate, becoming a racist, they will appreciate one and another even though have different colors and different spoken languages and customs.

For humans, when their shelter, food and healthcare is provided by robots and they do not need to worry about it, or fight to get it, then with proper education and thinking of a higher level of understanding all is left is to be more human, moral, and appreciate the world and environment they are living in and enjoy their surroundings nature, mountains, forest, deserts, rivers, oceans and all. They will be content and appreciate humans in wherever they are and whatever their specific livelihood and personality is with their own religion and way of life. There is no need for wars and segregation of people because of their colors and race.

CAPITALISM

Capitalism will change in the age of robotics. In the future due to increase in robot activities, most of the small businesses will be gone and only big businesses will thrive because of use of robots in lieu of high paying and costly workers. Therefore, the capitalism as we know will be changed. The big businesses will take over the small ones and become the norms in the society. We will see only few large companies providing all the necessary goods and services for the whole society. Small example of it is the Amazon, eBay, google etc. as we discussed earlier in this booklet, since driverless cars will do most of transportation needs of people and service companies, there would be a giant corporation such as Uber, Lift etc. who will take care of this matter.

Hence, there would be no need for people to buy cars or drive cars themselves when we used to have four or more cars in every household. By this, the capitalist notion of selling, leasing, and renting cars will be doomed. In such a system farming will be handled by robots therefore, there would not be any competition by farmers or wholesale companies since everything is per specified order. The small grocery stores with their limited item list will be the thing of the past because, the big companies will take over it. People rather walk to those places to buy groceries they prefer to order online, as we all experienced it at the time of pandemic. In the near future, giant delivery companies with their drones will do most of the grocery and other goods deliveries to the households, therefore, there would not be any competition between small delivery companies but will be few big companies who will be competing with each other rather than thousands of small companies.

In such a way the capitalism system as we know of, will be only active for big companies and not for the small businesses that is the reason that some are advocating "Universal Minimum Income," which is certain income paid to most of people due to unemployment. This money comes from taxation of those giant companies. It seems like in a new capitalism system, most people will live their life in a socialistic system but only the big companies will compete for bigger portion of the businesses for their country. Although, it is sad to see this segment of capitalist world is coming to the end for better or worse but millions of people who were working very hard in an irrational world of competition and trying to exceed others in sales and services now, they have to learn to enjoying a quieter life without much of competition and the stress.

We may think the quality of the goods will diminish because of lack of competition but since the big companies are taking over and robots are the ones who provide all the goods then the competition will be only within the big companies to increase their sells of similar goods with change in quality and quantities sold. The big companies to increase their sells need to provide higher quality and better designs for their goods, this intern will create competition between the producers of those goods. therefore, the managers and quality controllers of robots need to come up with a better and creative design of their items so online buyers will order more and companies could get more profit from the sales.

As we can see, the capitalist idea will survive in bigger companies' competition for more money and dominance. Due to creation of this new lifestyle for almost one third of the people who are staying home without working and getting average salary paid, those people may tend to be lazy and not seek for new jobs or education. They would be lazy soles staying home and wasting their time and energy on unnecessary items and smoking or drinking, giving their life away.

Also, this situation, may cause carelessness and breakdown of family because specially women could have many children by different partners as it happened in welfare societies in Midwest, years back. Not having scheduled life with requirements for providing for the family, some people tend to be lazy and use the easy root, be satisfied with whatever they have and being aloof will make them to be careless about their marriage and seek for satisfaction in other beds which intern breaks down marriages and ruin the existing family and due to those engagements new life is born with different partners which creates lots of pain and heartache for the kids and their families and many societal pains associated with this situation including depressions overuse of drugs and alcohol and ruining more lives.

LIFE IN COMMUNIST COUNTRIES

I have not lived in a communist country, but I have visited Armenia many times knowing that soviet Armenia used to be part of communist Soviet Union. Today, it is a free and democratic country. On my visits, I noticed few things of the past communist history. One of them is the multistory buildings, which are very similar. The other thing that I noticed is the backyard for those buildings are created such that the kids can play all the time by having water fountains and fruit trees nearby.

While the kids were playing, the elders from their apartment units were watching and keeping an eye on them. Sometimes, those elders would come down and teach the kids some tricks or tell stories about life and Armenian history or the world in general. Also, I noticed many parks full of kids and people playing around, listening to music, or getting foods and coffee from cafés from inside the parks; there were also pools or waterfalls in the park and many more amenities for people to enjoy their rest in the park. Another aspect of the life in Armenia is their love of music, arts, theater, cinemas, and their easy lifestyle at evenings and nights, gathering at outside cafés having coffee and drinks and listening to music while their kids are having fun. playing with other kids in the open field, staying late and enjoying one another.

In the near future, when robotic life will be upon us and there would be much more unemployment and people will be staying home, it would be a good idea to look at the lifestyle created by those communist counties to provide safety and enjoyment for the whole family outside their living quarters.

NEW WORLD SYSTEM

Robotics implemented will be the dawn of a new world system. A social system, which will change the way we lived for centuries and will bring upon us a new age of living with robots when the robots will do the work and thinking for us and we will live in harmony with nature as our great-grandparents did centuries ago. Basically, we will be living in heaven on earth when everything will be provided for us, including all life comforts, shelter, health, food, entertainment and all… the human beings will be living in heavenly life with no worries for work or providing sustenance for the family. All they must do is enjoy life, family, and get all the entertainment on the spot, no worries for sicknesses or pandemics because it all would be taken care of by robotic doctors and nurses.

Money is no issue anymore because there is no purpose for the money, humans are all equal and they all can get whatever their heart desires, including entertainment, free vacations, any kind of foods and drinks, live like a king with robots to serve them. In such a utopian life, people will be educated and trained in a higher mode and fulfill, God's desire to be honest, strive for the best, enjoy life and prayer, do the best they can without any outside requirements but their own. The school teaching will be only about true knowledge and well-being of humans and Godly works to enhance our devotion to God and our pure aspiration toward the true heaven coming at the end of their life.

Sarcastically, there is always jokes about heaven that when we are there what would we do? Sit around all day in the nature and enjoy heavenly life or there would be something else. In robotics heaven on earth, the people will have the time and health to enjoy all that God has provided them in nature, including all kinds of sports, arts, theater, movies, parties, family gatherings, Las Vegas games all day, fishing and hunting for those type of enthusiasts, and drinks and smoking for the bunch who love it. There would be no need for wars, (as Beatles sing about it) no reason for wars because everything is provided. There is no need to conquer other lands and kill people to acquire their riches because everything is free and a person is free to travel wherever he/she wants and live wherever he/she is comfortable. The language is not a barrier since the robots inside your brain will speak and translate in all kinds of languages in the world. You would know about those people's history, culture, language, habits, lifestyle, and feelings right away because your robotic mind will let you know instantly.

By the way, you want to travel? No worries! The best travel packages will be provided for you by use of robot personal airplanes with speeds beyond imagination. The world will be under your feet in few minutes or hours, wherever you wish to be. Human beings will not live only for hundred years, but they could live for two hundred or three hundred years due to advances in medicines and health provided by highly technical robot researchers and doctors. One of the best things about robots who do all that work for human beings is that they have no feelings, they do not get upset, or tired, or frowned upon; they do not get angry, or depressed.

Because of all mentioned above, there would be no fights between them or them and human beings. They would not be upset, because of the looks and talks of the owners. They will not be discouraged by upsetting words of human beings or they will not be encouraged or bullied either. On the other hand, knowing how the human beings are, with their emotions and feelings, they may train those robots to do crimes to kill, steal or create havoc to another human beings just to satisfy their owner's wish.

But due to abundance of life sources, I believe those fights will be minimum and will not be as World War III scenarios. It is only war of personal grievances in between friends or relatives in a positive thinking scenario. On the other hand, knowing the racist feelings among some nations, it would be possible that certain nations or its government would decide to eliminate other races. This scenario will cause huge wars between races. Also, the countries who have more robots and have more robotic armies may decide to take over other countries or the whole world as Hitler tried to do.

Robotics is an in-depth look at the new age we are on the verge of and, soon will be an age that is totally different than what human beings have experience before. The COVID-19 pandemic showed a glimpse of what the future is going to look like and how people will conduct their life and business affairs from inside their homes. Now, with the advent of new and highly technical robots who do most of the jobs for human beings, we will create a new world in which people do not need to work and will be travel by automated, robot-driven vehicles. In this insightful work, author Oksen Babakhanian and Ailin Babakhanian examine how people will survive and what will change about our society and culture. They believe the social life of the new age will be one of an "Educated Socialism" society.

About the Author

Oksen Babakhanian is an accomplished structural engineer, who owned his consulting structural engineering business for more than thirty years. He has recently started writing a few mini books after the sale of his business, pursuing his childhood dream of being a writer. His booklets are mostly about his experiences and world knowledge. As an engineer, he tried to enhance the social and economic situation of the society both at home and abroad. His works include *Rebirth of Cultural and Business Revolution in Armenia, Social Standing of African American People, Robotics,* and *Educated Socialism System* and *Economy of Waste*.

The author's first book; *The Economy of Waste*, has been published by Author house on 3/08/21. This book is about "Healthy" and "Unhealthy" economy and the balance between "Needs" and "wants" of people.

His second book; *The Pilgrimage: African American's Rebirth* was published by Author house on 4/16/21. This book is to revolutionize African American culture and mindset of slavery to a freedom that comes from the heart and mind and flow to reality of life, a life of freedom and respect for the soul of the African Americans.

www.ingramcontent.com/pod-product-compliance
Lightning Source LLC
Chambersburg PA
CBHW061520180526
45171CB00001B/262